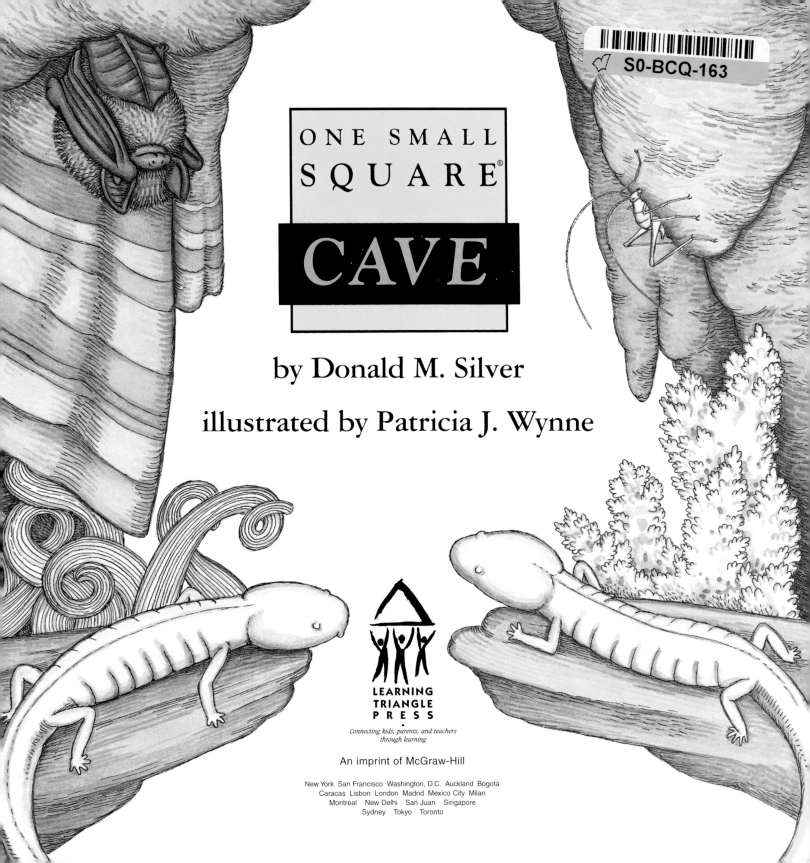

ONE SMALL SQUARE®

CAVE

by Donald M. Silver

illustrated by Patricia J. Wynne

LEARNING
TRIANGLE
PRESS

Connecting kids, parents, and teachers
through learning

An imprint of McGraw-Hill

New York San Francisco Washington, D.C. Auckland Bogotá
Caracas Lisbon London Madrid Mexico City Milan
Montreal New Delhi San Juan Singapore
Sydney Tokyo Toronto

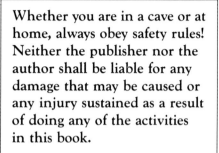

Whether you are in a cave or at home, always obey safety rules! Neither the publisher nor the author shall be liable for any damage that may be caused or any injury sustained as a result of doing any of the activities in this book.

Every plant and animal pictured in this book can be found with its name on pages 40–43. If you come to a word you don't know or can't pronounce, look for it on pages 45–47. The small diagram of a square on some pages shows the distance above the cave floor for that section of the book.

For Alf Hunter

—for taking such good care of Marge and being a super granddad to Daniel and Siobhan.

Thanks to Thomas L. Cathey for his insights into crystals; Maceo Mitchell for spelunking; Karen Malkus and Jeanne Gurnee for their suggestions on cave safety and activities; and Chester Beck for an unforgettable cave tour.

Text copyright © 1993 Donald M. Silver.
Illustrations copyright © 1993 Patricia J. Wynne.
All rights reserved.
One Small Square® is the registered trademark of Donald M. Silver and Patricia J. Wynne.

Library of Congress Cataloging Number 97-074149
ISBN 0-07-057929-6
3 4 5 6 7 8 9 QPD/QPD 9 0 2 1

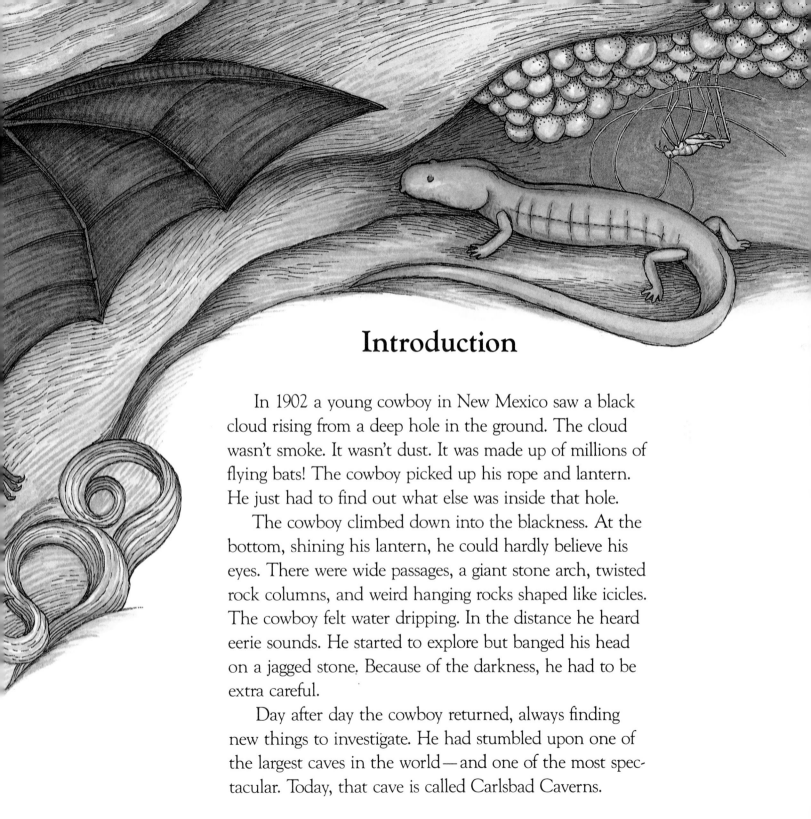

Introduction

In 1902 a young cowboy in New Mexico saw a black cloud rising from a deep hole in the ground. The cloud wasn't smoke. It wasn't dust. It was made up of millions of flying bats! The cowboy picked up his rope and lantern. He just had to find out what else was inside that hole.

The cowboy climbed down into the blackness. At the bottom, shining his lantern, he could hardly believe his eyes. There were wide passages, a giant stone arch, twisted rock columns, and weird hanging rocks shaped like icicles. The cowboy felt water dripping. In the distance he heard eerie sounds. He started to explore but banged his head on a jagged stone. Because of the darkness, he had to be extra careful.

Day after day the cowboy returned, always finding new things to investigate. He had stumbled upon one of the largest caves in the world—and one of the most spectacular. Today, that cave is called Carlsbad Caverns.

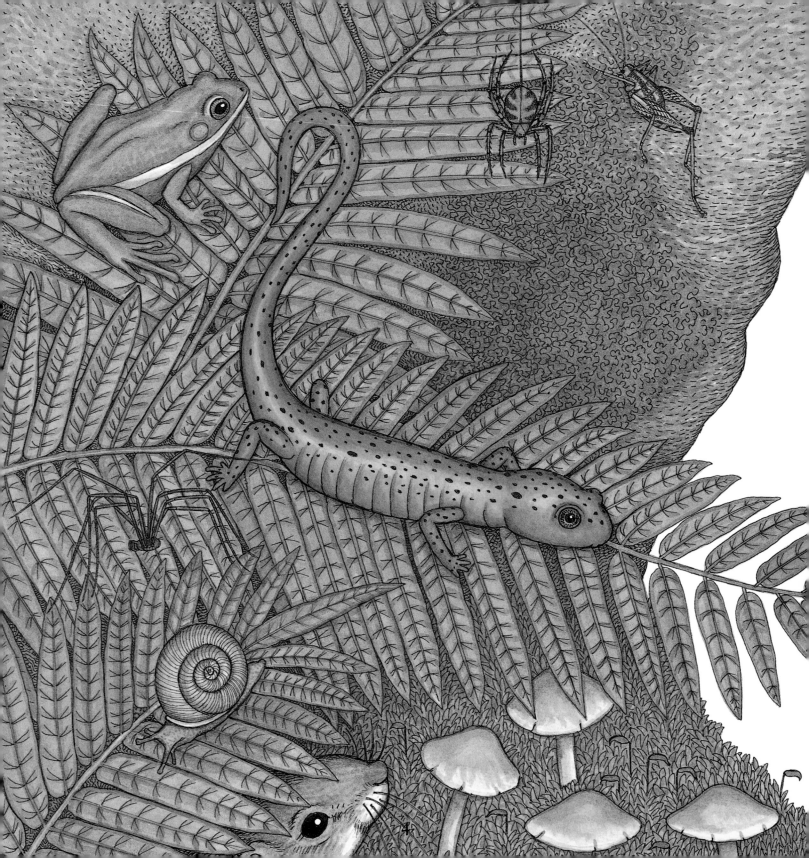

Unlike the cowboy, you must NEVER just climb down into any hole in the ground. The hole may collapse, or it may be so deep that you fall and hurt yourself. But that does not mean you cannot explore a cave.

Before entering a cave, be sure to ask an adult to help you find out who owns the land on which the cave is located. You must have the owner's permission to explore. An experienced adult must be with you at all times.

As the cowboy learned, a cave can be a dangerous place. For protection, you and the adult will need equipment like that shown on this page. Always tell another adult when you will be in the cave, its exact location, and when you expect to return. Try to wear a watch that beeps on the hour. It is so easy to lose track of time in a cave full of wonders.

Before you proceed, STOP! Check the SAFETY FIRST list on page 6 every time you visit a cave.

An inexpensive magnifying glass will let you get a close-up look at plants and animals without harming them. It will also let you see details in the rocks without touching them.

Wear a hard hat to protect your head (a bike helmet is fine); gloves to protect your hands; and boots or high sneakers with a good tread, so you don't slip on rocks. Carry a flashlight with spare batteries, a candle and matches in a waterproof container, and a plastic bottle filled with water.

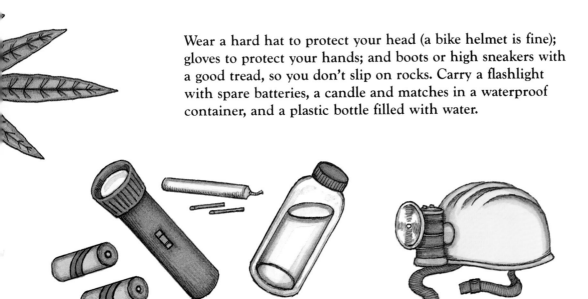

Safety First

1. NEVER enter a cave alone.
2. Explore a cave ONLY with an EXPERIENCED ADULT.
3. Get PERMISSION from the cave owner before you explore.
4. Make sure another adult knows WHERE you are going and WHEN you expect to return. Also leave a note with the same information on the windshield of the car you come in, if it is parked near the cave.
5. ALWAYS carry a FLASH-LIGHT with spare batteries and a CANDLE and MATCHES. If one light does not work, you'll have another to use.
6. NEVER go caving when it is raining or when it might rain. Caves can flood.
7. Wear WARM clothing unless your cave is located where the weather is warm all year round.
8. Ask the adult to CHECK OUT the cave before you enter.
9. NEVER run or jump in a cave.
10. NEVER drink cave water. You don't know what's in it.
11. NEVER touch cave animals or plants.
12. TAKE NOTHING from a cave. LEAVE NOTHING behind.

One Small Square at the Cave Entrance

Step into a cave and you enter a world of mystery. How did it come about? Why is it so spooky? What is creeping or crawling in the darkness? Has anyone else ever been inside?

You can solve these puzzles and discover how a cave works by exploring two small squares inside one. If you have a cave nearby, this book will tell you how. If not, after exploring the cave in the book, you will be prepared when you find a cave.

The first square is shown here. It is in the cave entrance, where some light reaches. Each side of this square is about as long as you are tall. The second square is on pages 16–17. It is in total darkness day and night.

As you follow along, there will be some activities you can do in the cave and some at home. None of the activities will harm the cave or its dwellers. Whenever you explore a cave, you must remember: Everything living in a cave struggles hard to survive. Harming any part of a cave can put wildlife in danger.

So check your flashlight batteries and get your equipment ready. If you look very closely, you may spot some of the creatures from this book in *your* cave. You may find yet others, depending on where your cave is. If you are lucky, you will discover a new kind of creature no one has ever seen.

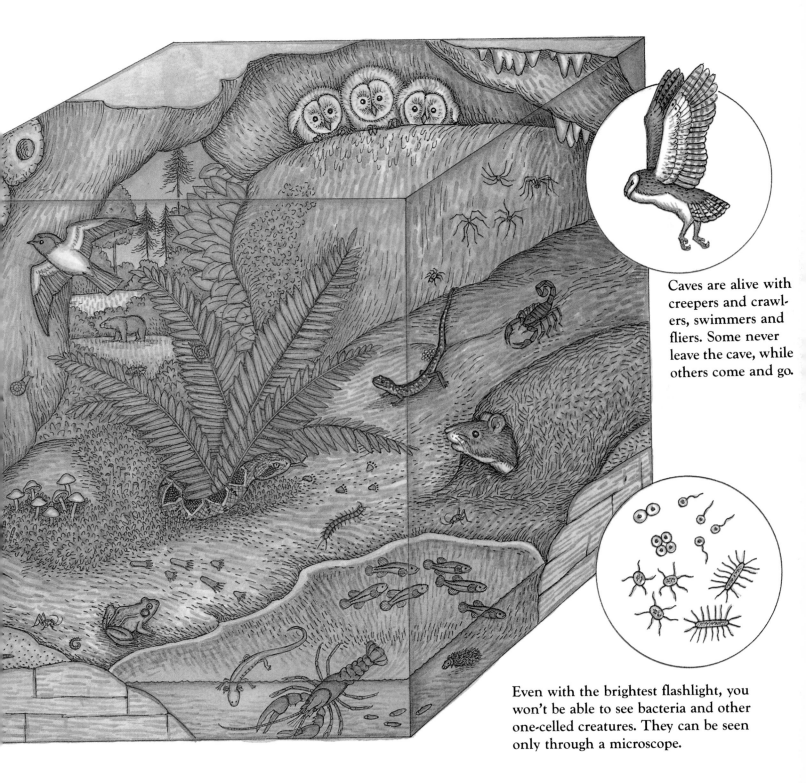

Caves are alive with creepers and crawlers, swimmers and fliers. Some never leave the cave, while others come and go.

Even with the brightest flashlight, you won't be able to see bacteria and other one-celled creatures. They can be seen only through a microscope.

7

Swiftly and silently a barn owl returns to the twilight zone with food for its young. You can tell a barn owl by its heart-shaped face.

Take a good look at any ferns, mosses, or wild-flowers in the twilight zone. You won't find them anywhere else in the cave. Like all other plants, they must capture light energy to make food.

The Twilight Zone

There is no bell to ring. There is no door at which to knock. Even so, it is a good idea to pause outside the opening of a cave. Look for animal footprints on the ground. A deer may have wandered into the cave, or a raccoon, or a skunk. So make some noise and step aside to give any animal visitors a chance to come out before you go in!

Flashlight in hand, you slowly enter a world without bright sunshine. With every step you move deeper into

8

On a cave wall in Southeast Asia swiftlets, like the one in the circle, build their nests out of saliva. Their saliva is different from yours. Thick and sticky, it hardens as it dries.

If you hear squeaks, cave swallows may be nesting above your square. Search for their mud nests with your light. The nests shelter eggs and young from harm. Phoebes (left) nest in caves too.

Your Cave Notebook

Always carry a notebook and a pen or a pencil when you explore a cave. Write down the date and the time you enter and leave the cave. Draw pictures of everything you find inside: animals, strange rock shapes, plants. Note where animals are hiding. Your notebook will help you remember what you saw so you can learn more about rocks and cave life later. It will turn into a guide to the wonders of the cave.

Using Your Senses

Sit in your small square and note what you can see, hear, and smell. Does this part of the cave feel warm or cool, dry or damp? Your senses will provide clues to how a cave works.

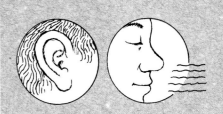

the twilight zone. This twilight zone is the only part of the cave touched by daylight. Still, it may be too dimly lit for you to see much. So switch on your flashlight when you need to.

You can choose any spot in the twilight zone for your small square, but try to include a cave wall or a stream as one edge. In those places you will discover more cave creatures than elsewhere.

If you find any plants at all in your cave, they will be in the twilight zone. These plants must have light to make food and grow. Just turn off your flashlight and you will see why the twilight zone isn't full of plants, like a backyard, a park, or a forest.

The twilight zone connects a cave with the outside world. If you live where there are heavy rains, part of your cave may flood. Before exploring, listen to the weather report. If there's a chance of rain, stay home.

Watch Your Step

You may want to begin exploring high up by lighting the ceiling. Or you may decide to inch along the cave wall and floor with your magnifying glass. Either way, search your small square for the animals of the twilight zone. Just watch your step. You never can tell what you will find—and where you will find it!

There may be hundreds of cave crickets wiggling their antennae as they scramble across the rocks. Or a wood rat sleeping inside a nest that it built of sticks. You may startle a red salamander hiding in a crack. Or cause a daddy longlegs to back away in order to escape your light beam.

Slimy salamander

Most salamanders lay their eggs in water. But not the slimy salamander. It lays them in damp caves and guards them until they hatch.

Amphipod

Copepod **Isopod**

Not all caves have streams that flow in from the outside. This one does. It is home to fishes, flatworms, crayfishes, and small animals related to lobsters and shrimps (see circle).

10

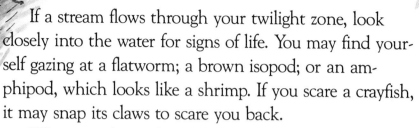

If a stream flows through your twilight zone, look closely into the water for signs of life. You may find yourself gazing at a flatworm; a brown isopod; or an amphipod, which looks like a shrimp. If you scare a crayfish, it may snap its claws to scare you back.

When you leave the twilight zone to return home, you won't be the only one making your way out of the cave. As night falls, the wood rat, the red salamander, and many of the crickets you saw earlier may head out to hunt for food. They will be joined by bats that take wing deeper inside the cave and quickly fly off to eat their evening meal. When you return in the daytime, the bats will be waiting for you to discover them in the blackness beyond the twilight zone.

A coin in your small square may be a clue that a wood rat is nesting there. The wood rat collects shiny or colorful objects when it hunts at night for food.

Open Wide

Walk into the cave and stand in your small square. How much can you see in the dim light? Keep looking. How much can you see five minutes later? Now shine your flashlight. How much more can you see?

At home, stand in a brightly lit room. Then take a look in a mirror at the black dot in each of your eyes. Now stand in a room with all the lights off for five minutes. Again look in a mirror. What happened to the black dots?

The black dots are the pupils of your eyes, the openings through which light enters your eyes. In bright light your pupils narrow to keep too much light from entering and harming the inner parts of your eyes. In dim light the pupils widen to let in as much light as possible, so you can see in places like the twilight zone.

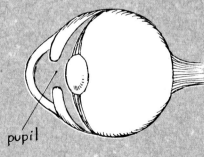

pupil

Cold air blowing into your cave can freeze water. Watch out for sharp icicles and slippery surfaces.

Use your magnifying glass to look for cutworm moths or a clump of daddy longlegs on the cave wall. These winter guests hardly move until the warm weather returns.

To escape the winter cold, garter snakes may enter your cave. They will look for a deep crack in the rocks and huddle together in a ball for the entire season.

Winter Guests

If you live where winters are cold, your small square will change as the days get shorter and shorter. Most of the plants die. Nesting birds fly off. Other twilight zone animals disappear as they move deeper inside the cave, where icy winds cannot reach them.

Will your small square become deserted? Perhaps not! Before the first frost, frogs, moths, bats, and snakes may check into your cave. So may a bear. All will become guests for the winter, safe from freezing and from predators—animals that would like to eat them.

Once inside your cave, each guest finds a spot that feels right for it. The spot may be in the twilight zone or in another part of the cave. The spot may be out in the open or in a nook or cranny.

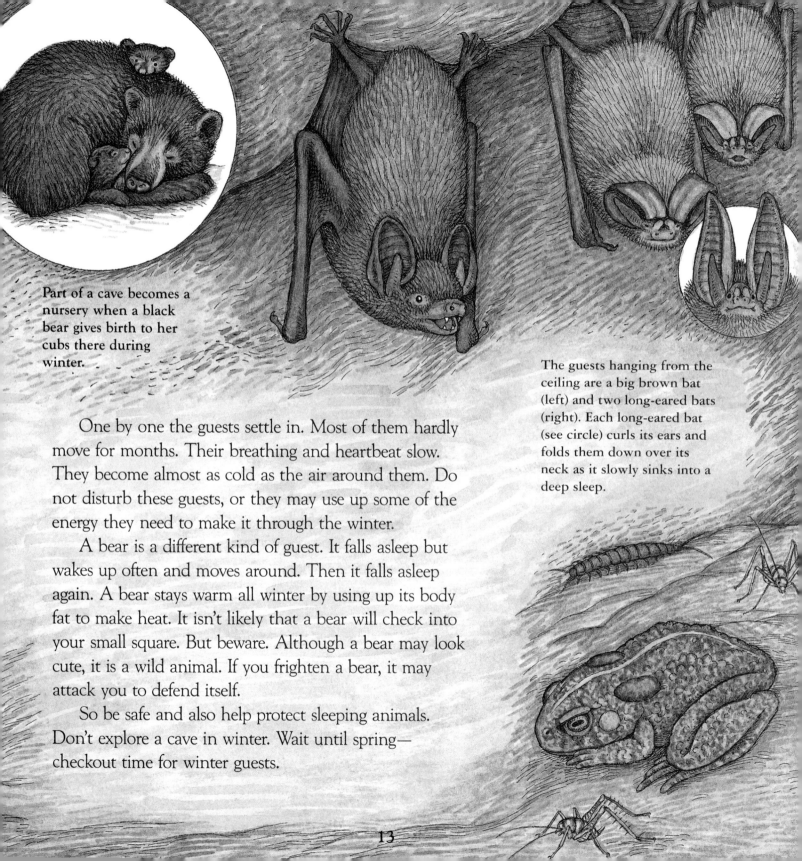

Part of a cave becomes a nursery when a black bear gives birth to her cubs there during winter.

The guests hanging from the ceiling are a big brown bat (left) and two long-eared bats (right). Each long-eared bat (see circle) curls its ears and folds them down over its neck as it slowly sinks into a deep sleep.

One by one the guests settle in. Most of them hardly move for months. Their breathing and heartbeat slow. They become almost as cold as the air around them. Do not disturb these guests, or they may use up some of the energy they need to make it through the winter.

A bear is a different kind of guest. It falls asleep but wakes up often and moves around. Then it falls asleep again. A bear stays warm all winter by using up its body fat to make heat. It isn't likely that a bear will check into your small square. But beware. Although a bear may look cute, it is a wild animal. If you frighten a bear, it may attack you to defend itself.

So be safe and also help protect sleeping animals. Don't explore a cave in winter. Wait until spring—checkout time for winter guests.

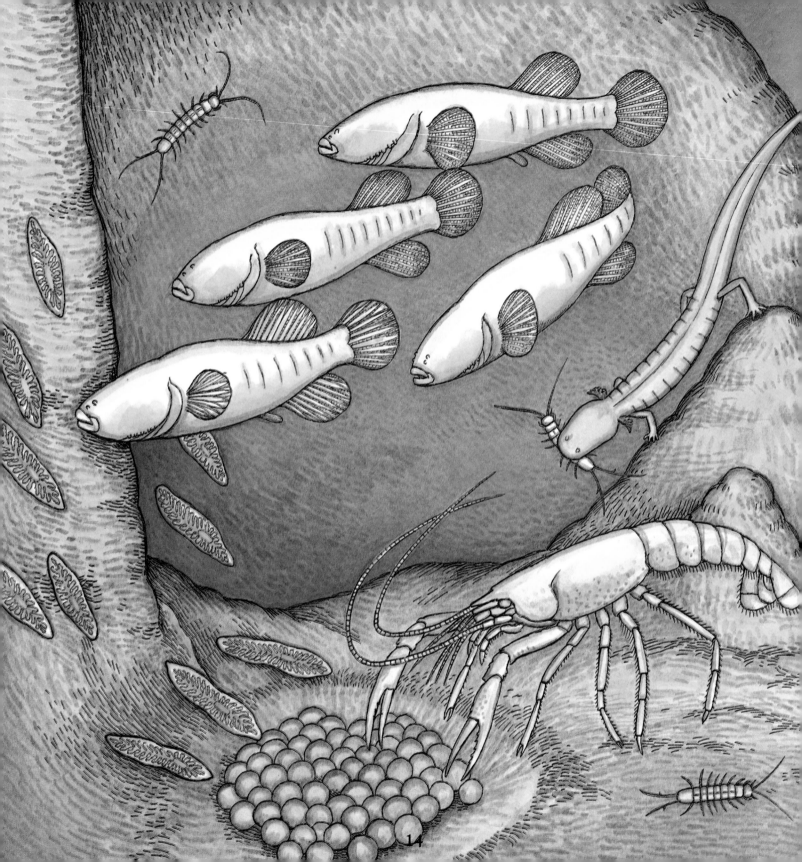

14

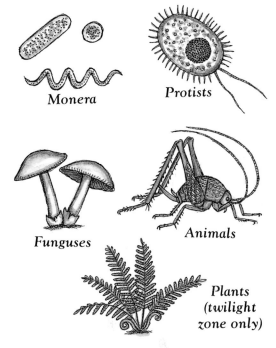

Monera

Protists

Funguses

Animals

Plants
(twilight
zone only)

One Small Square in Darkness

Beyond the twilight zone there is total darkness. You enter a world in which your eyes are useless without your flashlight. This sunless world can be very chilly and damp. It may smell musty—stale and moldy. It may also smell like ammonia, a strong cleaning liquid. Perhaps you will hear water dripping—or nothing at all. The stillness is eerie. Isn't anything crawling around in here?

Imagine trying to live in a place like this—with no trees, flowers, or sunlight. What would you eat? How would you stay warm? How long would your flashlight batteries last?

Your pets could not survive here. Nor could the squirrels, birds, butterflies, and bees you see in the park. But there are strange, ghostly cave creatures that cannot live anywhere else. This is their home.

Before you explore this part of the cave, STOP! Check the safety list on page 6 again. Here you can see only as far as your flashlight shines. Experienced cavers won't

Cave "Trogs"

Cave creatures can be divided into three groups:
1. Cave visitors—trogloxenes—such as bats live only part of their lives in caves.
2. Cave lovers—troglophiles—such as some kinds of spiders and crickets can live in caves or in places outside that are dark and damp like caves.
3. Cave dwellers—troglobites—such as blind cavefishes spend their entire lives in caves.

Cave visitors

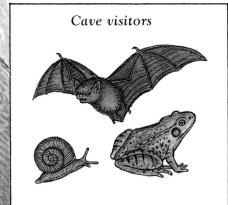

Cave lovers

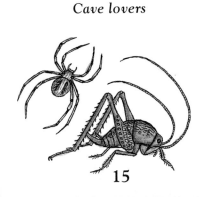

Cave dwellers

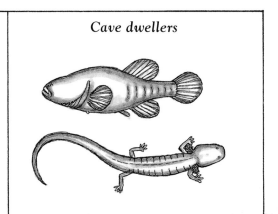

15

enter the dark part of a cave alone. They know how dangerous a cave can be when you can't see what lies ahead.

Choose a spot for your small square that is not too far beyond the end of the twilight zone. If a stream flows through this part of the cave, try to include it on one side of your square. The square illustrated on these pages is about the same size as the twilight zone small square. (This square will always be shown as if it were fully lit by a spotlight. Otherwise you could not see anything.)

The dark part of a cave is like nowhere else on Earth. But it is easily harmed. So shine your light, look through your magnifying glass, and fill your notebook. But don't touch the rocks or animals. Also, remember to carry out of the cave everything you brought in with you. Leave nothing behind but your footprints!

Your flashlight will show you that the dark part of a cave is one of nature's most beautiful sights. Its colorful stone treasures will dazzle your eyes.

16

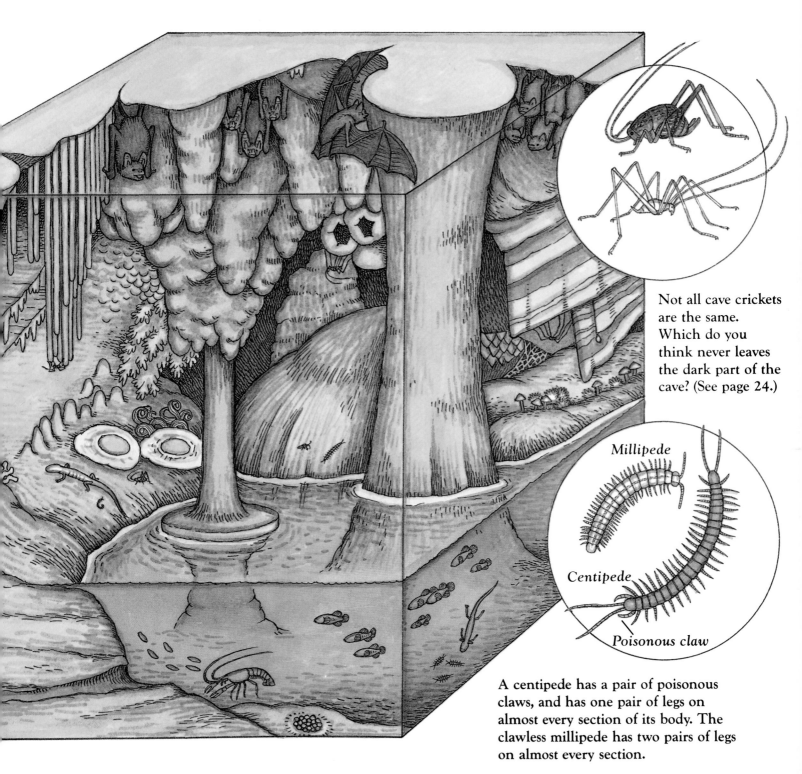

Not all cave crickets are the same. Which do you think never leaves the dark part of the cave? (See page 24.)

Millipede

Centipede

Poisonous claw

A centipede has a pair of poisonous claws, and has one pair of legs on almost every section of its body. The clawless millipede has two pairs of legs on almost every section.

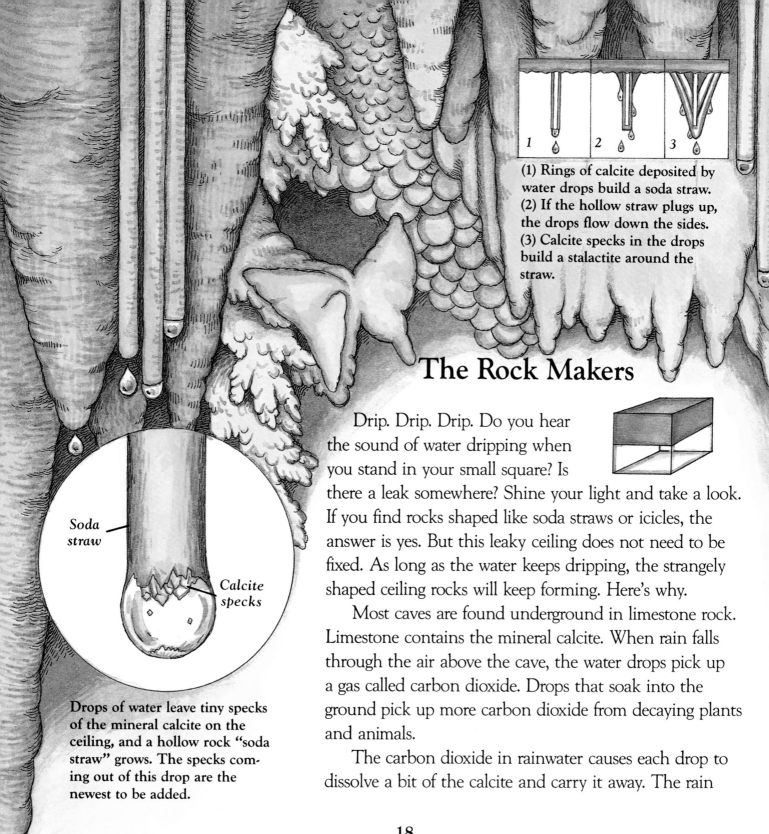

(1) Rings of calcite deposited by water drops build a soda straw.
(2) If the hollow straw plugs up, the drops flow down the sides.
(3) Calcite specks in the drops build a stalactite around the straw.

Soda straw

Calcite specks

Drops of water leave tiny specks of the mineral calcite on the ceiling, and a hollow rock "soda straw" grows. The specks coming out of this drop are the newest to be added.

The Rock Makers

Drip. Drip. Drip. Do you hear the sound of water dripping when you stand in your small square? Is there a leak somewhere? Shine your light and take a look. If you find rocks shaped like soda straws or icicles, the answer is yes. But this leaky ceiling does not need to be fixed. As long as the water keeps dripping, the strangely shaped ceiling rocks will keep forming. Here's why.

Most caves are found underground in limestone rock. Limestone contains the mineral calcite. When rain falls through the air above the cave, the water drops pick up a gas called carbon dioxide. Drops that soak into the ground pick up more carbon dioxide from decaying plants and animals.

The carbon dioxide in rainwater causes each drop to dissolve a bit of the calcite and carry it away. The rain

18

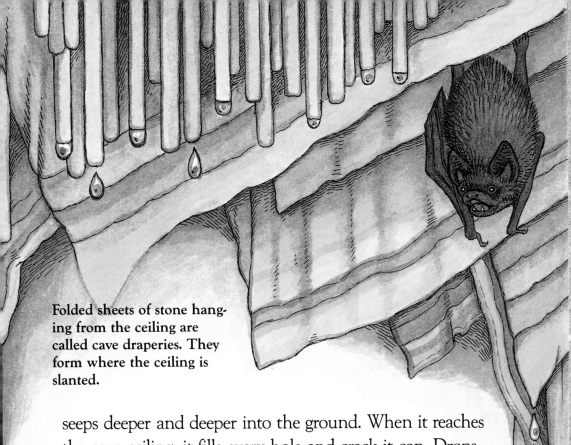

Folded sheets of stone hanging from the ceiling are called cave draperies. They form where the ceiling is slanted.

seeps deeper and deeper into the ground. When it reaches the cave ceiling, it fills every hole and crack it can. Drops form on the ceiling. In the open cave below, however, the carbon dioxide escapes. Without the carbon dioxide, the water cannot hold the calcite. A speck of calcite comes out of each drop and is deposited on the ceiling. The next drop in the same spot adds another speck on top of the first.

Speck by speck this calcite hardens into rocks that become shaped like soda straws. If the straws become clogged, the specks build icicle-shaped rocks called stalactites. Don't expect to see the stalactites grow. Most likely it took thousands of years and countless drops to create the ceiling formations above your small square.

The ceiling isn't the only part of the cave decorated by water drops. Shine your light on the floor and search for

Now You See It

You can't dissolve calcite at home, but you can dissolve table salt. Add about 15 shakes of salt into a cup of water. Stir until all the salt dissolves.

salt

water

Just as water dissolves salt, rain with added carbon dioxide dissolves calcite—only much slower.

Now fill a medicine dropper (or straw) with salty water. Hold the dropper about an inch above a flat dish and release ten drops, one at a time. Space the drops.

Let the dish sit until the drops disappear. This happens because the water evaporates—changes from a liquid to a gas. In your notebook draw what is left behind on the dish. It is easy now to see how specks of calcite build soda straws on a cave ceiling.

Repeat on a rainy day. How long does it take the water to evaporate? In a cave the air is often damp. Calcite comes out of drops when carbon dioxide is lost even before water evaporates.

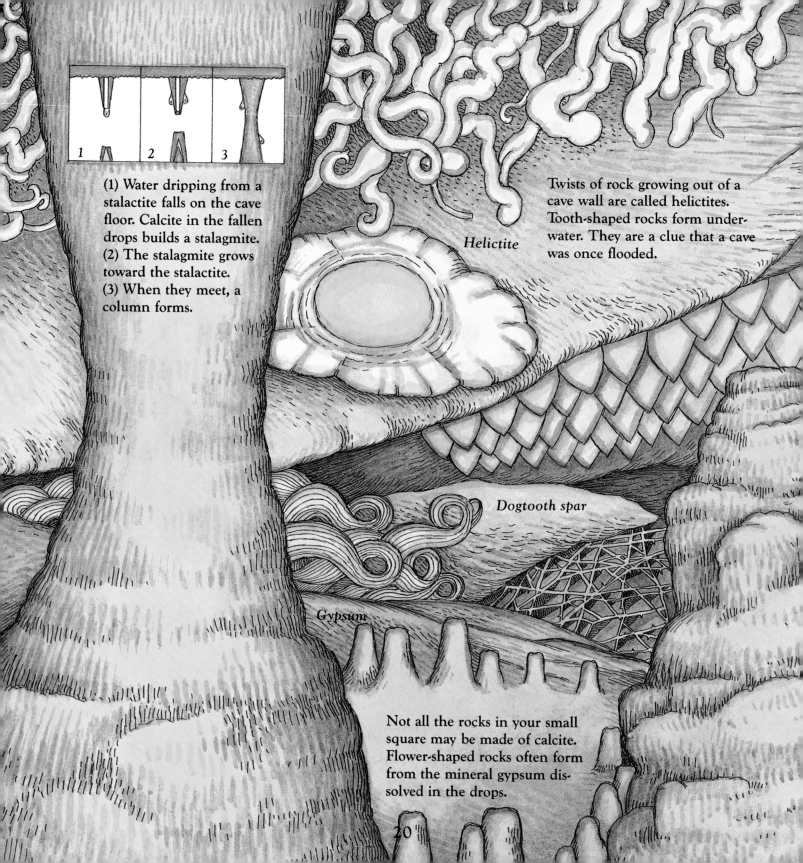

(1) Water dripping from a stalactite falls on the cave floor. Calcite in the fallen drops builds a stalagmite.
(2) The stalagmite grows toward the stalactite.
(3) When they meet, a column forms.

Twists of rock growing out of a cave wall are called helictites. Tooth-shaped rocks form underwater. They are a clue that a cave was once flooded.

Helictite

Dogtooth spar

Gypsum

Not all the rocks in your small square may be made of calcite. Flower-shaped rocks often form from the mineral gypsum dissolved in the drops.

20

rounded rocks called stalagmites. They formed gradually from specks of calcite in water that dripped to the cave floor from the ceiling or from the tips of stalactites.

Don't forget to check the cave walls, too, to view what wonders the water has worked. Draw every different shape in your notebook. Some of them may look very weird. Cavers around the world have discovered rocks shaped like slabs of bacon, grapes, butterflies, corn, folded drapes, flowers, twists, threads, and curls, just to mention a few. You can name the rocks in your small square by what they look like to you.

Besides being beautiful, stalactites, stalagmites, and other cave rocks form nooks and crannies where animals can rest and hide.

If explorers touched the cave formations, the rocks would soon break or turn black from the natural oil on the explorers' hands. And the cracks and the holes through which the water drips might seal up.

Drip. Drip. Drip. When cavers tell you a cave is alive, they don't mean only that there are creatures inside. They mean that the rock makers are still at work.

This cave was once flooded, or there was a pool of water here. How can you tell? Because shelfstone has formed. Thin sheets of water flowing over the floor built the flowstone.

Shelfstone

Flowstone

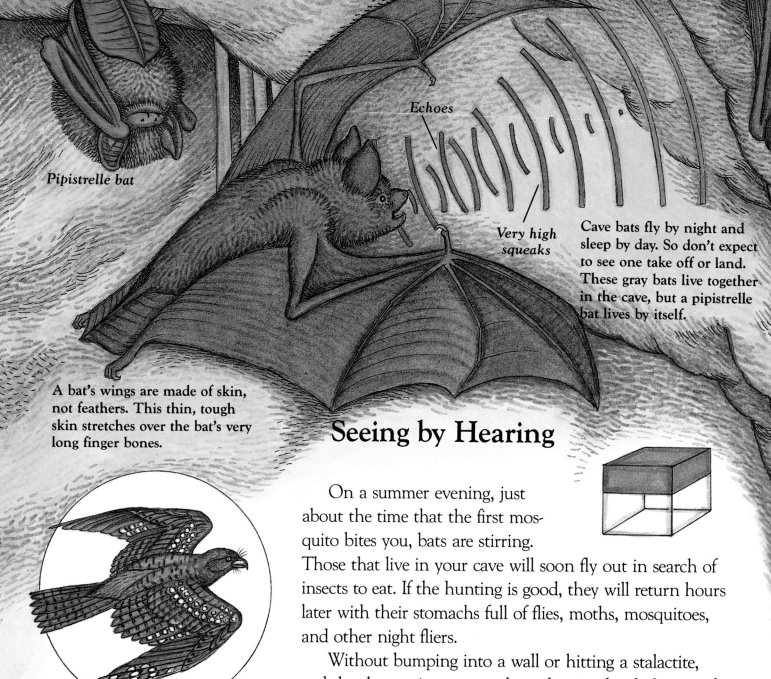

Pipistrelle bat

Echoes

Very high squeaks

A bat's wings are made of skin, not feathers. This thin, tough skin stretches over the bat's very long finger bones.

Cave bats fly by night and sleep by day. So don't expect to see one take off or land. These gray bats live together in the cave, but a pipistrelle bat lives by itself.

Like bats, oilbirds use sounds and echoes to fly through total darkness. They live deep inside caves in northern South America and on Trinidad, an island off its coast.

Seeing by Hearing

On a summer evening, just about the time that the first mosquito bites you, bats are stirring. Those that live in your cave will soon fly out in search of insects to eat. If the hunting is good, they will return hours later with their stomachs full of flies, moths, mosquitoes, and other night fliers.

Without bumping into a wall or hitting a stalactite, each bat locates its spot on the ceiling in the dark part of the cave. One by one the bats hook their curved toe claws onto rock. If any bats land above your small square, you will find them hanging there upside down, sleeping the day away.

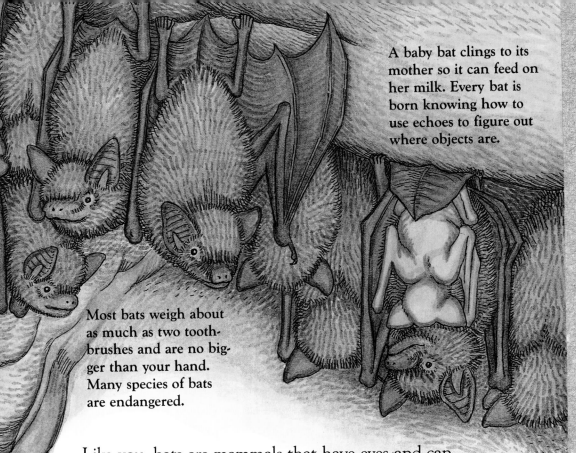

A baby bat clings to its mother so it can feed on her milk. Every bat is born knowing how to use echoes to figure out where objects are.

Most bats weigh about as much as two toothbrushes and are no bigger than your hand. Many species of bats are endangered.

Like you, bats are mammals that have eyes and can see. Like you, bats cannot see a thing where there is no light. But they don't need a flashlight to pass unharmed through the maze of stalactites and stalagmites. Unlike you, bats can "see" by hearing.

As a bat flies, it sends out very high squeaks that people cannot hear. The squeaks bounce off everything they hit and echo back to the bat's ears. The echoes tell the bats the size, the shape, and the position of stalactites, stalagmites, cave walls, and other bats. Outside the cave the echoes pinpoint flying insects so the bat can swoop down and capture them.

Bats are gentle. They will not attack you. But let them sleep. A bat that is touched will be frightened that a predator is near, and it may bite you.

Listen Closely

Why do bats have big ears? Make a funnel out of paper and place it over one of your ears. Which ear hears better now?

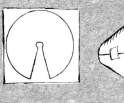

Echo......
Echo...Echo

The next time you are in a place where your voice echoes, shout out your name and you will hear it repeated. Ask a friend to call out at the same time you do. What happens? Isn't it amazing that thousands of bats, each making sounds, can fly out of a cave in the evening without being confused by all the echoes?

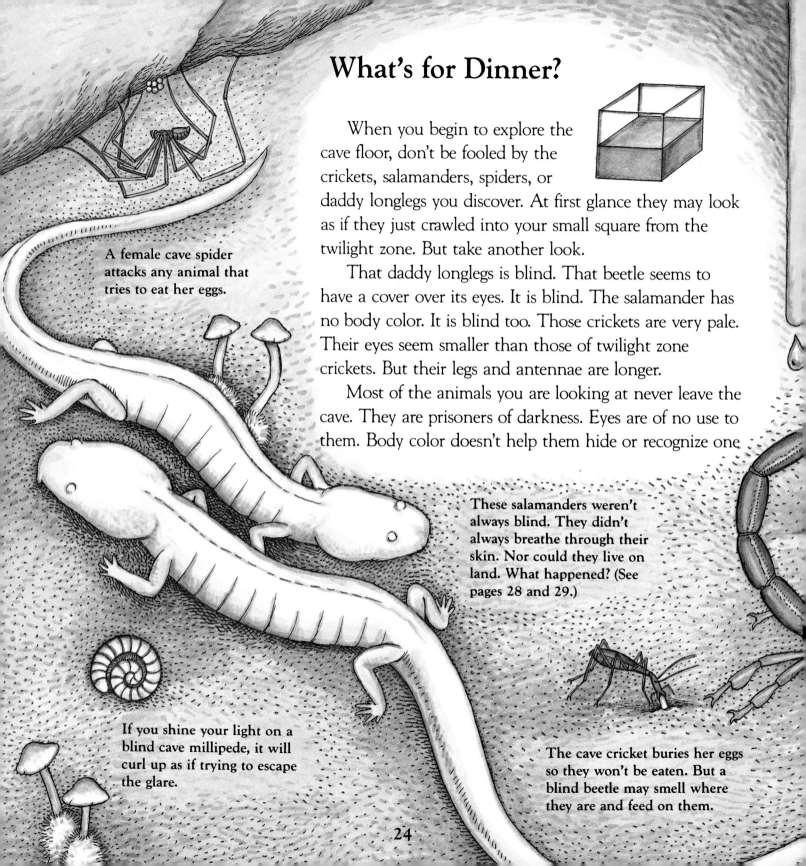

What's for Dinner?

When you begin to explore the cave floor, don't be fooled by the crickets, salamanders, spiders, or daddy longlegs you discover. At first glance they may look as if they just crawled into your small square from the twilight zone. But take another look.

That daddy longlegs is blind. That beetle seems to have a cover over its eyes. It is blind. The salamander has no body color. It is blind too. Those crickets are very pale. Their eyes seem smaller than those of twilight zone crickets. But their legs and antennae are longer.

Most of the animals you are looking at never leave the cave. They are prisoners of darkness. Eyes are of no use to them. Body color doesn't help them hide or recognize one

A female cave spider attacks any animal that tries to eat her eggs.

These salamanders weren't always blind. They didn't always breathe through their skin. Nor could they live on land. What happened? (See pages 28 and 29.)

If you shine your light on a blind cave millipede, it will curl up as if trying to escape the glare.

The cave cricket buries her eggs so they won't be eaten. But a blind beetle may smell where they are and feed on them.

another. Long antennae, though, can feel for food and sense if anything is moving nearby. Long legs can cover more distance with fewer steps, which saves the insect energy. And the less energy these creatures use, the better. Energy comes from food, and food can be very hard to find in the dark part of a cave.

Outside caves, on land and sea, plants capture energy from the sun and use it to make food. Plant eaters feed on them, and meat eaters eat the plant eaters. In this way energy from the sun passes from plants to animals, to be used for breathing, growing, moving, and escaping danger.

In the dark part of a cave there are no plants to eat. There seem to be few animals to eat either. If there are more, they are very well hidden.

With the tip of its tail, a scorpion can sting a cave cricket and turn it into a meal. Be careful not to let a scorpion sting you.

Above, a daddy longlegs dines on a cricket leg. Below, mold growing on a dead cricket becomes food for mold eaters. Molds sometimes make caves smell musty.

The Same? Or Not?

In your notebook draw each animal you see living on the cave floor. Compare these pictures with those you drew of the twilight zone animals. How are they alike? How are they different?

Now look through field guides, books that help you identify insects, birds, rocks—just about everything. Start anywhere—say, with salamanders. Do cave salamanders look the same as the salamanders living outside the cave? What about beetles? Crickets? You can draw these animals in your notebook too.

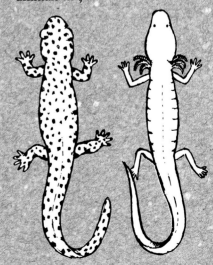

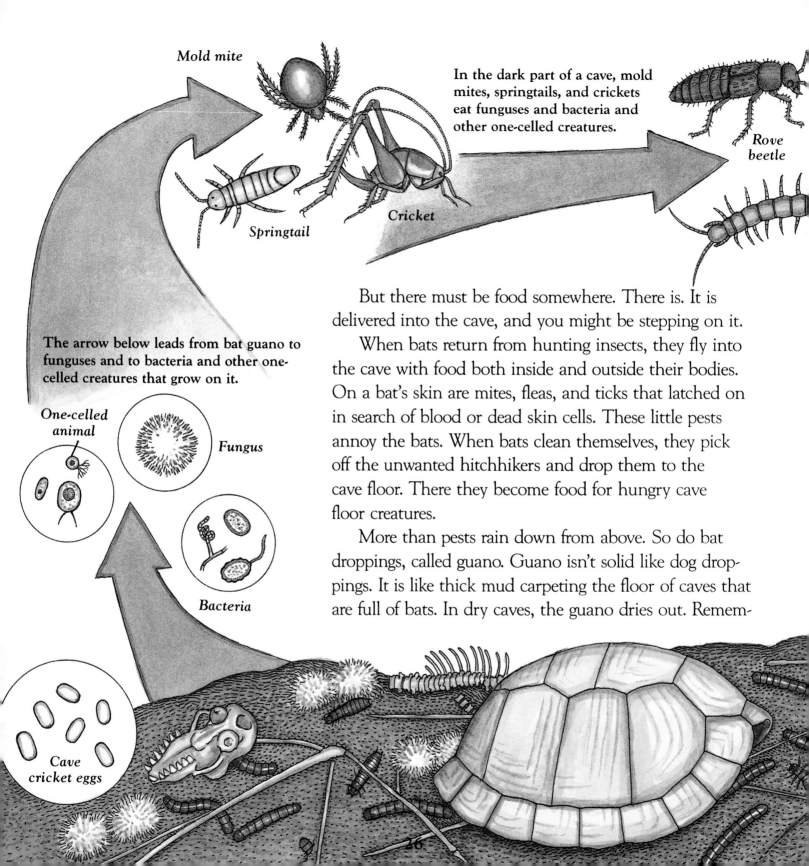

Mold mite

In the dark part of a cave, mold mites, springtails, and crickets eat funguses and bacteria and other one-celled creatures.

Rove beetle

Springtail

Cricket

The arrow below leads from bat guano to funguses and to bacteria and other one-celled creatures that grow on it.

One-celled animal

Fungus

Bacteria

Cave cricket eggs

But there must be food somewhere. There is. It is delivered into the cave, and you might be stepping on it.

When bats return from hunting insects, they fly into the cave with food both inside and outside their bodies. On a bat's skin are mites, fleas, and ticks that latched on in search of blood or dead skin cells. These little pests annoy the bats. When bats clean themselves, they pick off the unwanted hitchhikers and drop them to the cave floor. There they become food for hungry cave floor creatures.

More than pests rain down from above. So do bat droppings, called guano. Guano isn't solid like dog droppings. It is like thick mud carpeting the floor of caves that are full of bats. In dry caves, the guano dries out. Remem-

Cave mite

Centipede

Snail

Both arrows leading to the rove beetle, the cave mite, and the centipede come from creatures these animals eat. The arrow at the bottom right goes from guano to guano eaters.

Guano beetle

Millipede

Beetle larva

Roundworm

ber the strong smell of ammonia? It comes from guano.

Guano contains food that has passed through bats' bodies. Funguses and bacteria and other one-celled creatures grow on guano. They, too, become food. If there is guano in your small square, look at it under your magnifying glass. You may see worms, springtails, and beetles feeding on it or on white "hairs" sticking up like those on stale bread. The "hairs" are funguses.

Bats aren't the only cave animals to step out for dinner. On warm, damp nights cave crickets leave to feed on plants near the cave entrance. When they return, they add their droppings to the cave menu.

The animals in your small square can't afford to be fussy eaters. Nothing is wasted.

When bats die, their bodies fall to the cave floor and are eaten down to the bones. So are the bodies of turtles or raccoons that wander into the cave and die.

27

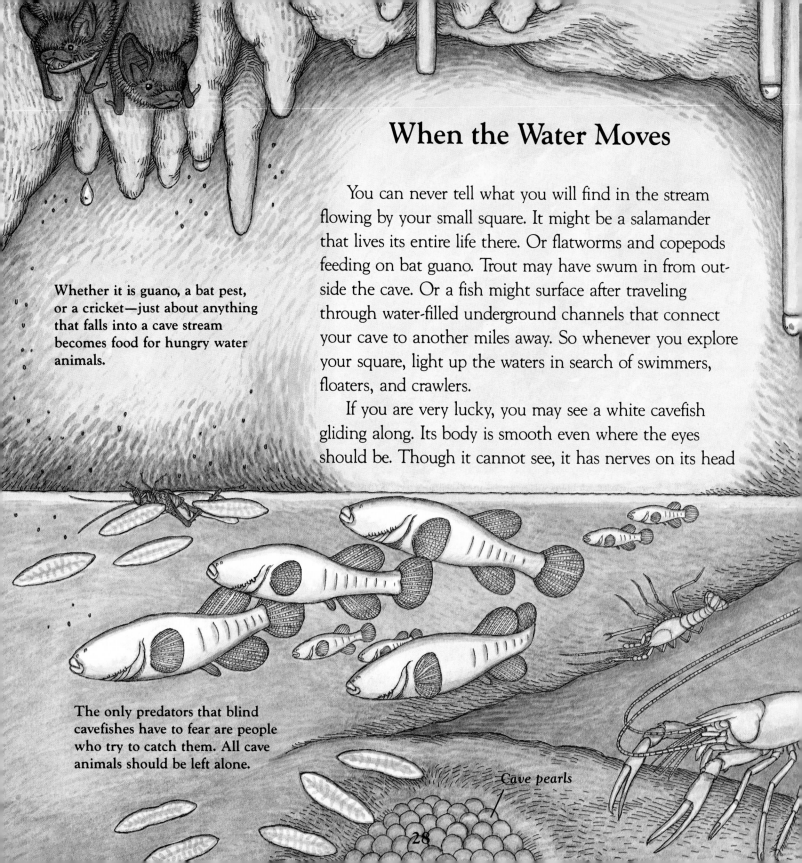

When the Water Moves

You can never tell what you will find in the stream flowing by your small square. It might be a salamander that lives its entire life there. Or flatworms and copepods feeding on bat guano. Trout may have swum in from outside the cave. Or a fish might surface after traveling through water-filled underground channels that connect your cave to another miles away. So whenever you explore your square, light up the waters in search of swimmers, floaters, and crawlers.

If you are very lucky, you may see a white cavefish gliding along. Its body is smooth even where the eyes should be. Though it cannot see, it has nerves on its head

Whether it is guano, a bat pest, or a cricket—just about anything that falls into a cave stream becomes food for hungry water animals.

The only predators that blind cavefishes have to fear are people who try to catch them. All cave animals should be left alone.

Cave pearls

28

and sides that feel the slightest movement in the water. The motion can lead the cavefish to its next meal— a tasty blind isopod or amphipod, perhaps.

Sometimes the movements lead to a baby fish. Watch closely, for as soon as the baby senses the water moving nearby, it freezes in place. Often this is enough to save the baby from being eaten by mistake.

Your light may also fall on a young, dark salamander with small black eyes and feathery red gills for breathing. If you could observe the salamander turn into an adult, you would see its gills disappear, its color lighten, and its eyelids grow together, never to reopen. Soon the blind salamander would crawl out of the water to live on the cave floor, breathing through its thin skin.

Proteus

Texas cave salamander

Not all cave salamanders lose their eyes and gills when they turn into adults as Ozark blind salamanders do. Proteus and the Texas cave salamander are born blind and keep their gills for life. Proteus lives in caves in Europe.

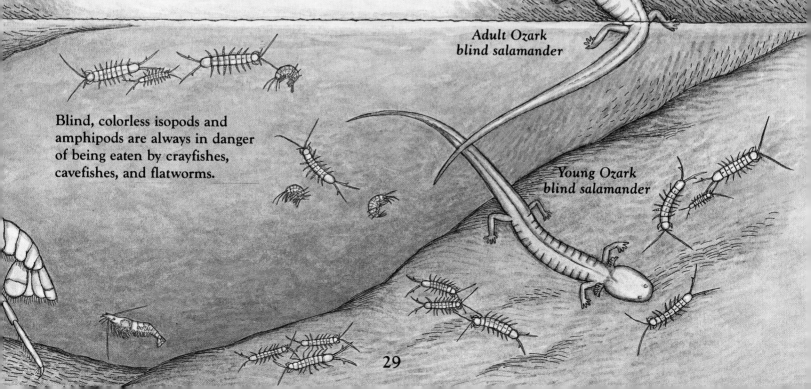

Adult Ozark blind salamander

Blind, colorless isopods and amphipods are always in danger of being eaten by crayfishes, cavefishes, and flatworms.

Young Ozark blind salamander

Cave in a Box

Take a shoebox and measure its length and depth. Cut a piece of paper for the background cave wall about a quarter-inch less deep than the box but about four inches longer. Draw and color rocks on it. You can include drawings of animals like those shown on page 36. Place the picture in the shoebox and tape each side to the front. The picture will curve.

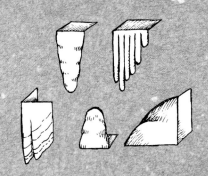

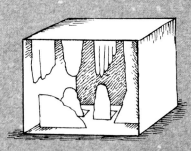

On a separate sheet, draw rocks of different shapes, each with a flap at the side. Cut each out, bend its flap, and glue it to the sides of the box. Draw, color, and cut out stalactites (flaps at the top) and stalagmites (flaps at the bottom). Glue stalactites to the ceiling and stalagmites to the floor of your cave in a box.

The crayfishes in this part of the cave are blind and white too. Led by their long antennae, they crawl slowly along the stream bottom. By "tasting" chemicals in the water, a blind crayfish can tell where there are bacteria, flatworms, or isopods to eat.

Streams deliver bacteria, one-celled animals, leaves, and bits and pieces of other kinds of food into caves. In some places, melting snow and heavy spring rains flood caves. The floodwaters wash in twigs, bark, seeds, leaves, and small animals. For a time there is a lot more to eat, but never as much as in the world outside.

Keeping Caves Alive

Once you begin to explore a cave, you may not want to stop. At every twist and turn there is something new: rock shapes you never imagined, crystals that sparkle like jewels, possibly an underground river. Only in a cave can you marvel at how beetles, crickets, crayfishes, and salamanders find food and stay alive in complete darkness.

But to go any farther than your small square, you will need a guide who has explored the cave before or someone trained in what to do should you get lost, slip on mud, fall down a hole, or find yourself unable to continue exploring.

Adult Ozark blind cave salamander

Blind millipede

Pipistrelle bat

Cave butterfly

Helictites

Scorpion

Camel-backed cave cricket

Eyeless flatworm

Small Square in Darkness

Can you match each living thing and rock formation to its outline?

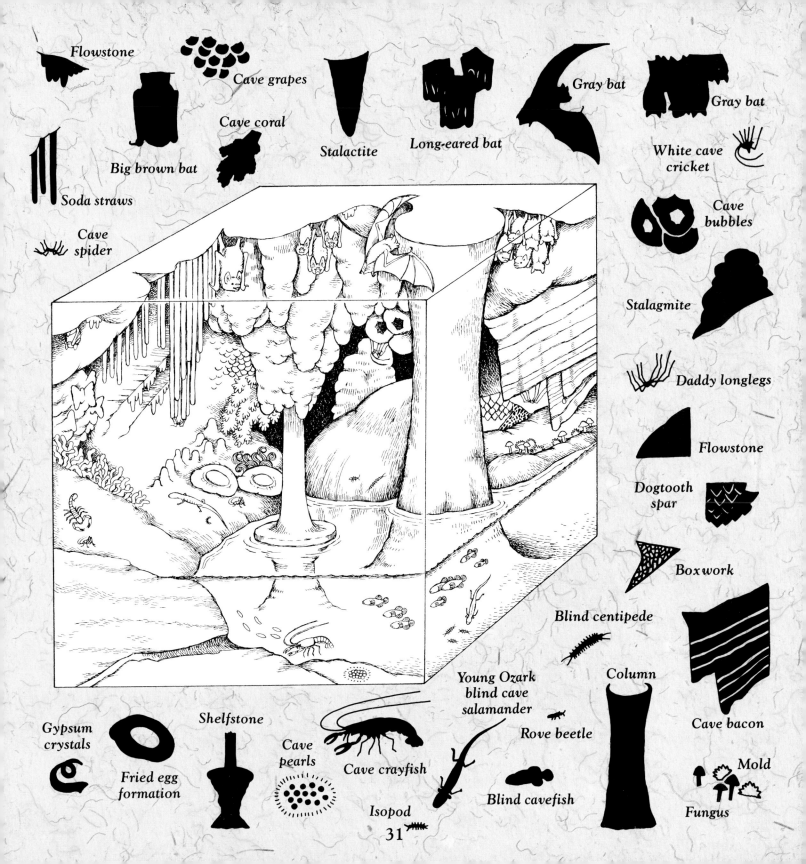

Flowstone

Cave grapes

Cave coral

Stalactite

Long-eared bat

Gray bat

Gray bat

White cave cricket

Big brown bat

Soda straws

Cave spider

Cave bubbles

Stalagmite

Daddy longlegs

Flowstone

Dogtooth spar

Boxwork

Blind centipede

Cave bacon

Column

Gypsum crystals

Shelfstone

Young Ozark blind cave salamander

Rove beetle

Fried egg formation

Cave pearls

Cave crayfish

Blind cavefish

Mold

Fungus

Isopod

31

When the Wind Blows

Is air moving through your cave? Hold a piece of paper between your thumb and first finger so it hangs as shown.

If the paper moves toward the dark part of the cave, air is moving in. If the paper moves toward the twilight zone, air is flowing out. If a cave has more than one opening, air will blow right through. Try this activity in the twilight zone and in the dark part of your cave, on a calm day and on a windy one, in spring and in autumn.

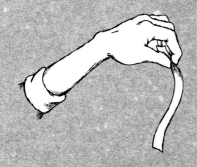

Tell from Spel

When you see *spel* at the start of a word, the word is likely to have something to do with caves. Many English words come from ancient Greek. The Greek word for cave was *spelaion*.

Scientists who study caves are speleologists. Cave explorers like you are spelunkers (although they like to be called cavers). Stalactites, stalagmites, and other cave formations are speleothems.

Everyone who enters a cave should wear a hard hat. Cavers put headlamps on their hard hats so they can leave their arms and hands free for easier climbing and exploring (see page 44). They are careful not to disturb bats and other animals that are sleeping through the winter. And they never explore a cave alone—NEVER.

Every year thousands of women and men enter the mysterious underground world of caves to gaze on hidden wonders of nature. They study how rocks form. They search for clues to the past (see page 37). And they try to discover how each cave creature is so well fitted, or adapted, to survive.

There is still much to be learned from caves. But many caves and cave creatures are in danger of being destroyed by pollution from garbage, poisonous chemicals, and other wastes from cities and factories. Too often caves are also permanently damaged by people who don't do their best to be safe cavers.

You can help keep caves alive by harming nothing inside a cave, taking nothing from a cave, leaving nothing behind, and telling an adult if you see or smell chemicals or wastes near a cave. Caring about the rocks and the animals is the best way to make sure that caves will be around for a long, long time.

Cave snail

Mushroom

Red cave salamander

Millipede

Cave cricket

Rove beetle

Moss

Earthworm

Green frog

Raccoon tracks

Green algae

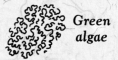

Small Square at Cave Entrance

Can you match each living thing to its outline?

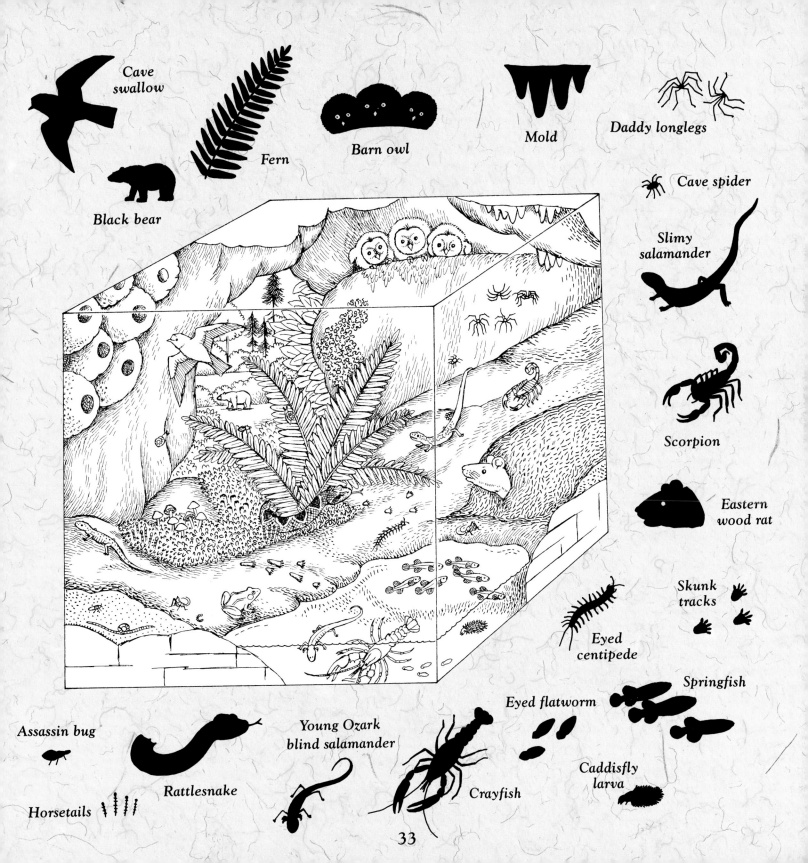

Cave swallow

Fern

Barn owl

Mold

Daddy longlegs

Black bear

Cave spider

Slimy salamander

Scorpion

Eastern wood rat

Skunk tracks

Eyed centipede

Springfish

Eyed flatworm

Assassin bug

Young Ozark blind salamander

Caddisfly larva

Rattlesnake

Crayfish

Horsetails

33

The Cave Makers

Water does more than decorate a cave with unusual rocks. It helps make caves. Follow along to find out how.

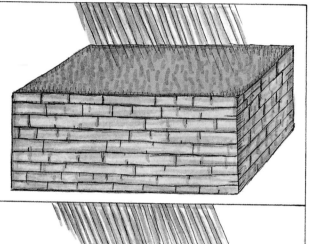

1. Once this cave was solid limestone. Rain picked up carbon dioxide from the air and soil and seeped into tiny cracks in the limestone. It started to dissolve the mineral calcite in the rock. Very slowly the cracks grew into holes.

2. Water kept filling the underground holes and dissolving more calcite. The holes grew into larger and larger cavities, and the water seeped deeper and deeper into the rock. Over thousands of years the cavities connected. They formed the passages, chambers, and shafts of a cave.

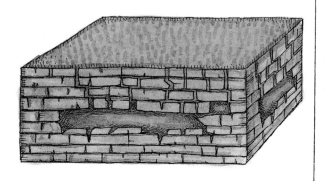

3. The underground water flowed from the passages, chambers, and shafts. But water kept dripping from cracks in the cave ceiling and walls. Stalactites and stalagmites started to grow. They are still forming.

34

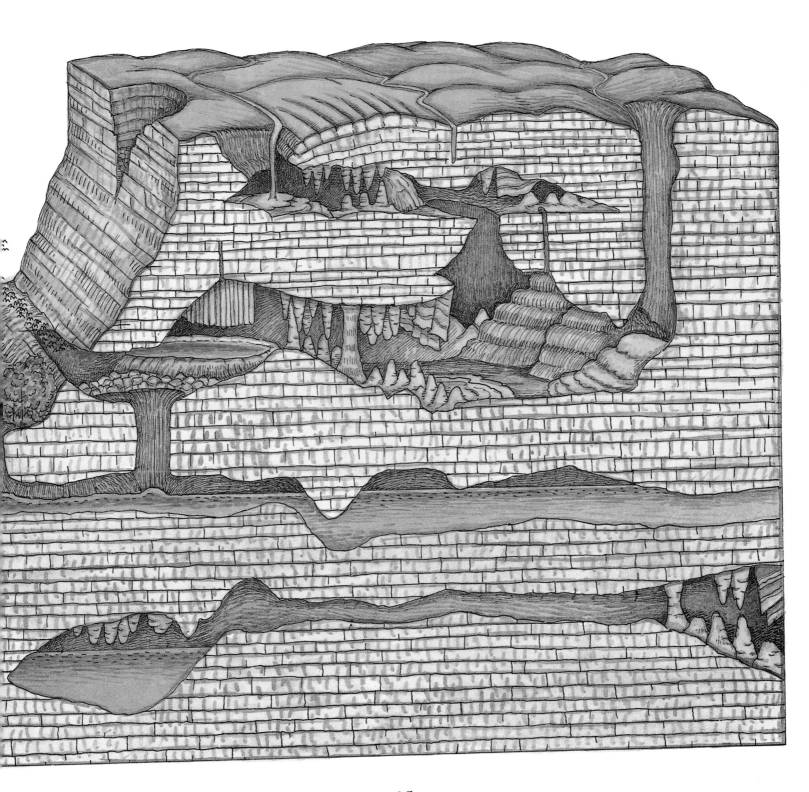

Caves of the World

One day you may get to visit one or all of these limestone caves found around the world. Each year new caves are discovered by cavers like you.

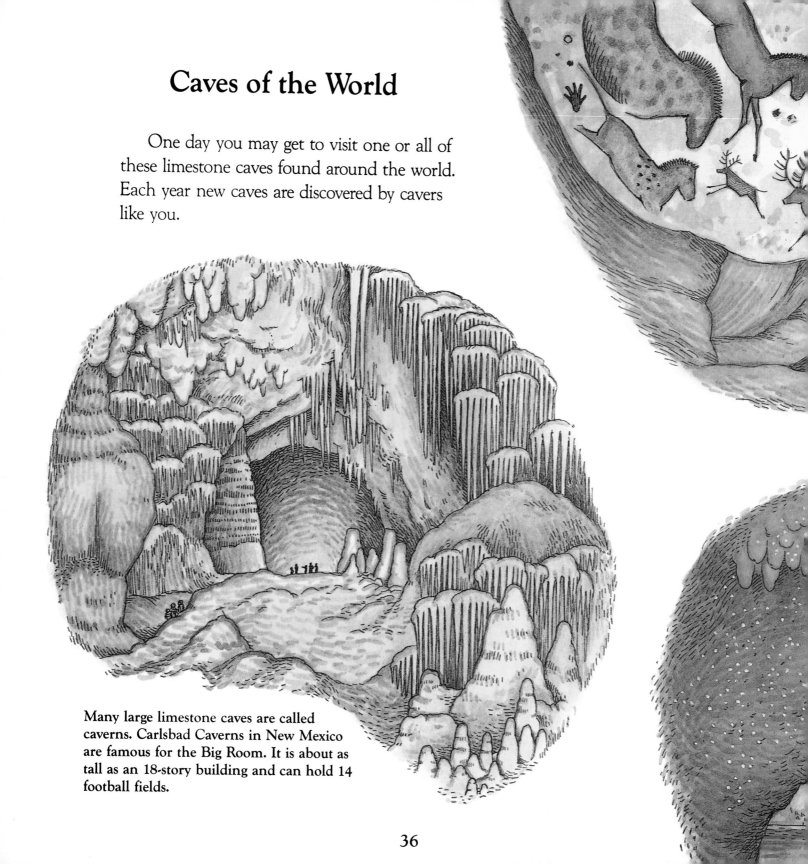

Many large limestone caves are called caverns. Carlsbad Caverns in New Mexico are famous for the Big Room. It is about as tall as an 18-story building and can hold 14 football fields.

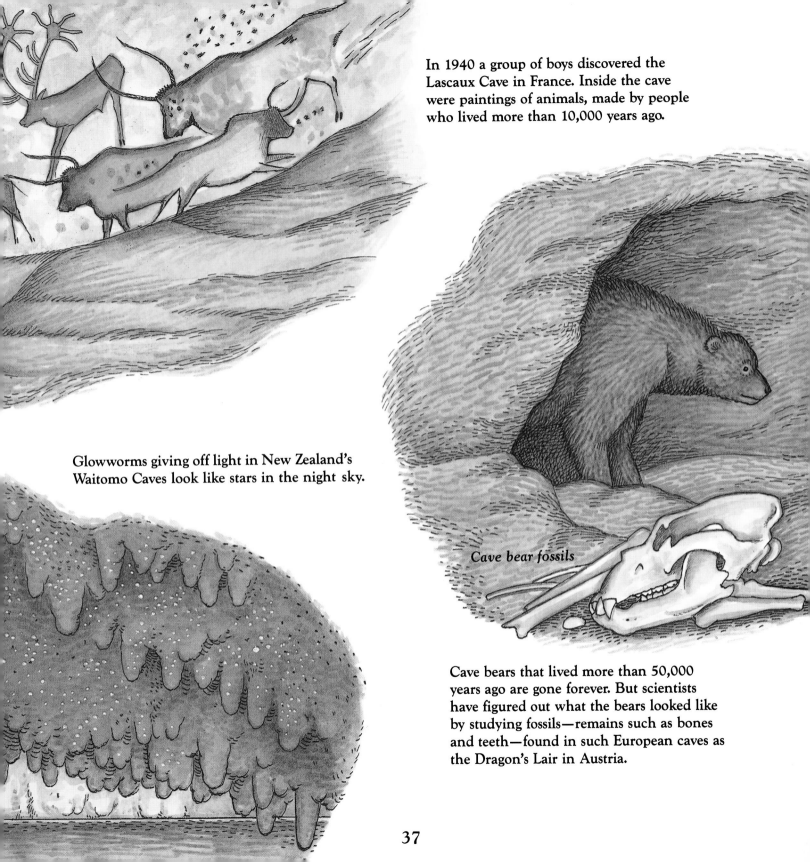

In 1940 a group of boys discovered the Lascaux Cave in France. Inside the cave were paintings of animals, made by people who lived more than 10,000 years ago.

Glowworms giving off light in New Zealand's Waitomo Caves look like stars in the night sky.

Cave bear fossils

Cave bears that lived more than 50,000 years ago are gone forever. But scientists have figured out what the bears looked like by studying fossils—remains such as bones and teeth—found in such European caves as the Dragon's Lair in Austria.

37

Other Kinds of Caves

Not all caves are found in limestone. Nature hollows out other types of caves where there are glaciers, sea cliffs, or volcanoes.

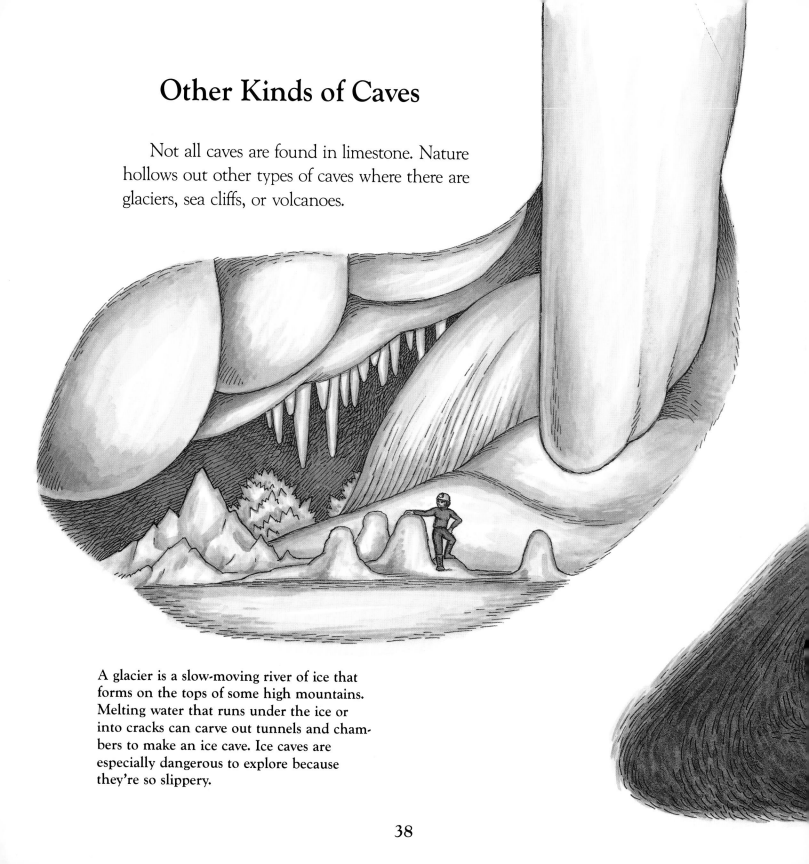

A glacier is a slow-moving river of ice that forms on the tops of some high mountains. Melting water that runs under the ice or into cracks can carve out tunnels and chambers to make an ice cave. Ice caves are especially dangerous to explore because they're so slippery.

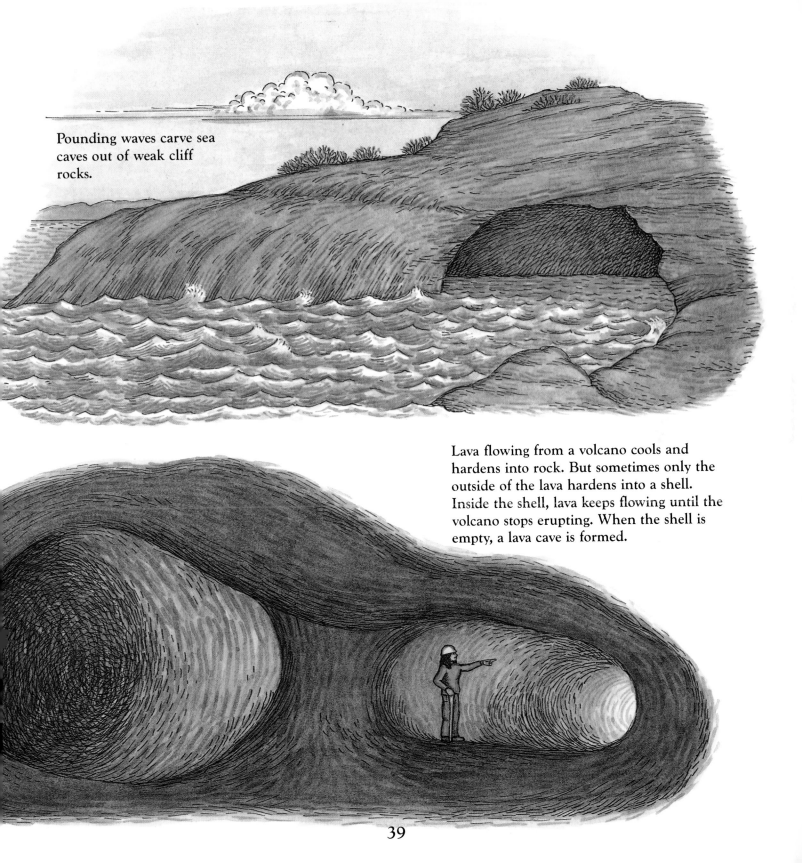

Pounding waves carve sea caves out of weak cliff rocks.

Lava flowing from a volcano cools and hardens into rock. But sometimes only the outside of the lava hardens into a shell. Inside the shell, lava keeps flowing until the volcano stops erupting. When the shell is empty, a lava cave is formed.

All these cave animals are vertebrates. They have bones in their bodies.

Look for fur on mammals; feathers on birds; dry scales on reptiles; tough scales on most fishes; and a thin, scaleless skin on amphibians.

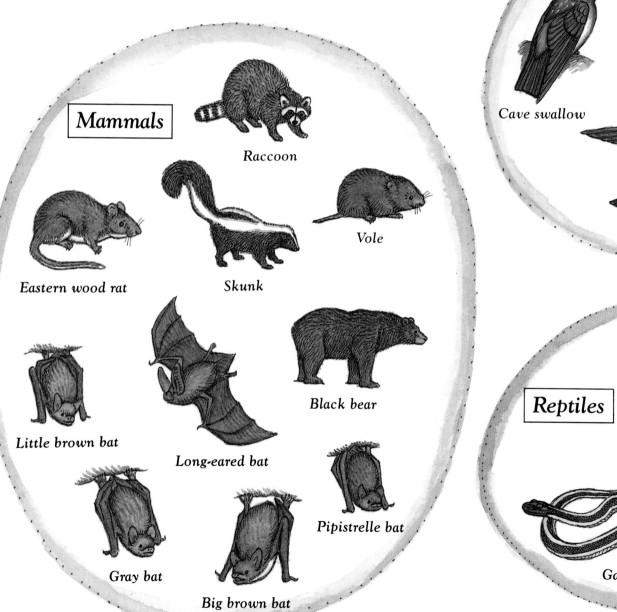

Birds

Cave swallow

Eastern phoebe

Cave swiftlet

Mammals

Raccoon

Eastern wood rat

Skunk

Vole

Little brown bat

Long-eared bat

Black bear

Pipistrelle bat

Gray bat

Big brown bat

Reptiles

Rattlesnake

Garter snake

40

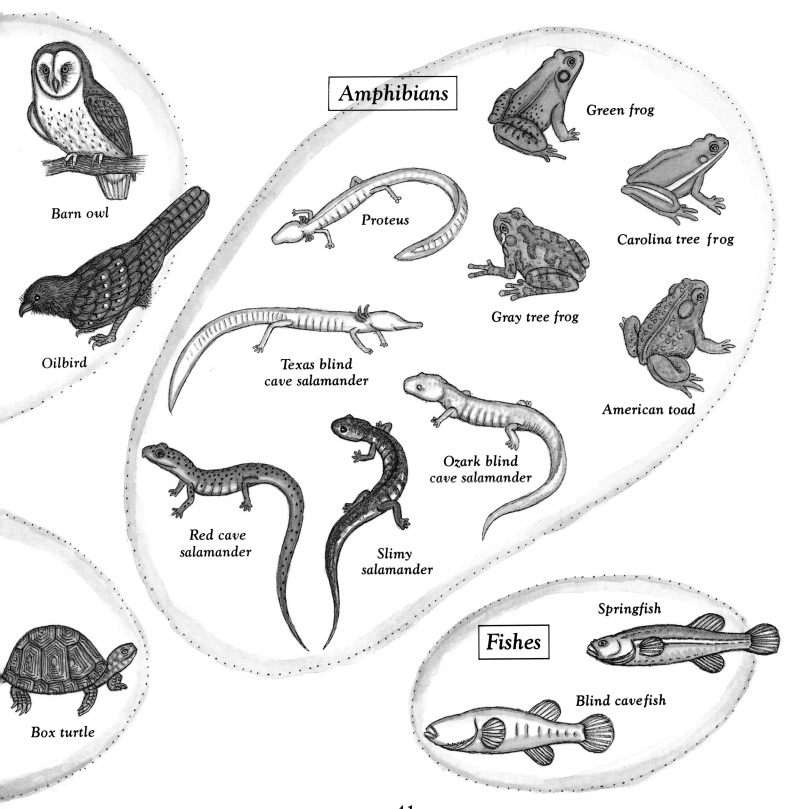

Barn owl

Oilbird

Box turtle

Amphibians

Green frog

Proteus

Carolina tree frog

Gray tree frog

American toad

Texas blind
cave salamander

Ozark blind
cave salamander

Red cave
salamander

Slimy
salamander

Fishes

Springfish

Blind cavefish

41

These cave animals are invertebrates. They have no bones. Except for the worms and snails, all of them have jointed legs and hard crusts covering their soft bodies.

Be sure to examine plants, funguses, rocks, and minerals under your magnifying glass. To see one-celled monera and protists, you would need a microscope.

Invertebrates

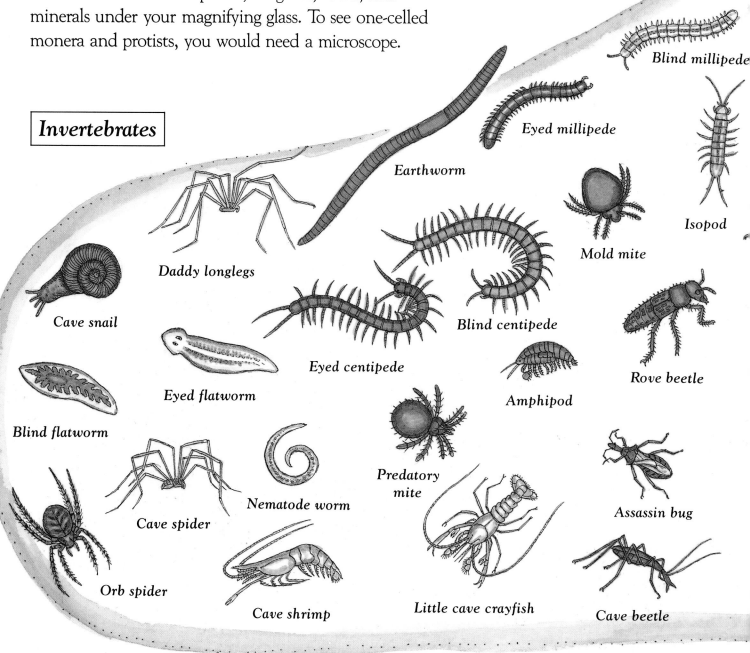

Blind millipede

Eyed millipede

Earthworm

Isopod

Mold mite

Cave snail

Daddy longlegs

Blind centipede

Rove beetle

Eyed centipede

Eyed flatworm

Amphipod

Blind flatworm

Predatory mite

Cave spider

Nematode worm

Assassin bug

Orb spider

Cave shrimp

Little cave crayfish

Cave beetle

42

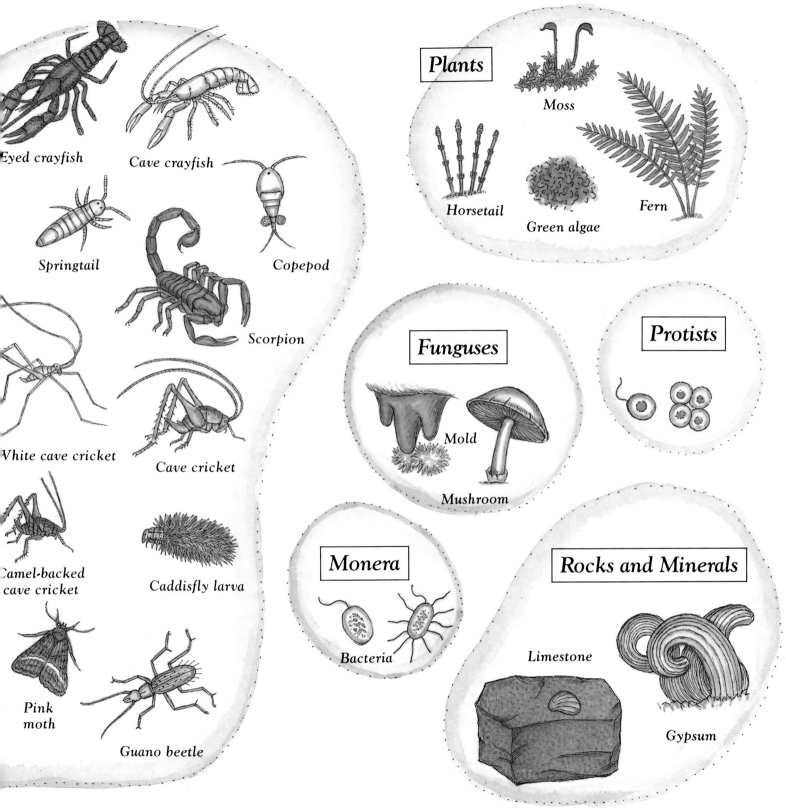

Eyed crayfish

Cave crayfish

Springtail

Copepod

Scorpion

White cave cricket

Cave cricket

Camel-backed cave cricket

Caddisfly larva

Pink moth

Guano beetle

Plants

Moss

Horsetail

Green algae

Fern

Funguses

Mold

Mushroom

Protists

Monera

Bacteria

Rocks and Minerals

Limestone

Gypsum

43

The Safe Caver

Caving can be very dangerous. Cavers must be well trained to climb up and down and to select and use equipment like that shown here. They must also learn when to *stop* exploring. Cavers return to the surface if they begin to get tired or if the way ahead looks too difficult. They know it is harder to get back up than it is to go down.

Hard hat with carbide
or electric lamp

Strong rope

First-aid kit

Blankets

Backpack with water
and high-energy food

Hand-operated ascender
(helps a caver climb a rope)

Rappel rack (controls how fast
a caver drops down a rope)

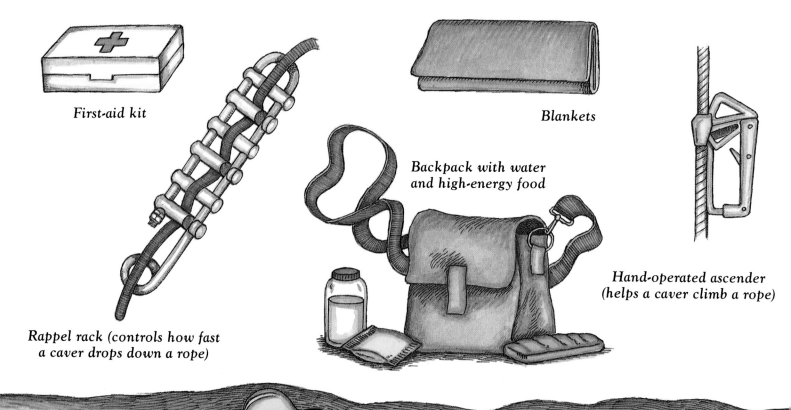

Index

A
adaptation 32. *Any part of a living thing that makes it fitted to survive where it lives.*
ammonia 15, 27
amphibian (am-FIB-ee-in) 40. *Bony animal that lives the first part of its life in water and the second part on land.*

amphipod (AM-fuh-pahd) 11, 29
animal 5, 6, 8, 9, 10, 13, 15, 16, 18, 21, 25, 28, 30, 32, 37
antenna 10, 24, 25, 30

B
bacteria 7, 26, 27, 30. *Kinds of monera—one-celled creatures that don't have a nucleus.*
bat 3, 11, 12, 13, 22, 23, 26, 27, 28, 32
bear 12, 13
beetle 24, 25, 27, 30
bird 12, 40
blindness 24, 29, 30

C
calcite 18, 19, 20, 21, 34

carbon dioxide 18, 19, 34
Carlsbad Caverns 3, 36
cave bear 37
cave draperies 18
cavefish 15, 28, 29
cave paintings 37
caver 16, 21, 32, 36, 44
cavern 36
cell 7, 26, 27, 30, 42. *Smallest living part of all plants, animals, and funguses. Some living things are made up of just one cell.*
centipede 17, 27
claw 11, 17, 22
cliff 38, 39
column 20
copepod (KOH-puh-pahd) 28
crayfish 10, 11, 29, 30
cricket 10, 11, 15, 17, 24, 25, 27, 28, 30
crystal 30

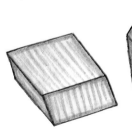

D
daddy longlegs 10, 12, 24, 25
dampness 9, 15
darkness 3, 6, 15, 22, 24, 30
deer 8
Dragon's Lair 37
droppings 26, 27

E
echo 22, 23

egg 9, 10, 24
endangered species 22
energy 13, 25. *Ability to do work or to cause changes.*
evaporation 19
eye 11, 23, 24, 28

F
fern 8
field guide 25
fish 10, 28, 29, 40
flashlight 5, 6, 7, 8, 11, 15, 16, 23
flatworm 10, 11, 28, 29, 30
flea 26
flowstone 21
footprint 8
formation 19, 21. *Shape of rock or rocks.*
fossil 37. *Anything that remains of a plant or animal that lived a long time ago.*

frog 12
frost 12
fungus 15, 26, 27, 42

Index

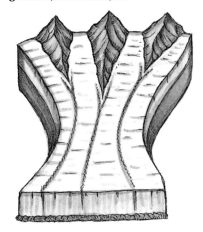

Index

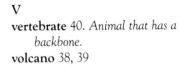

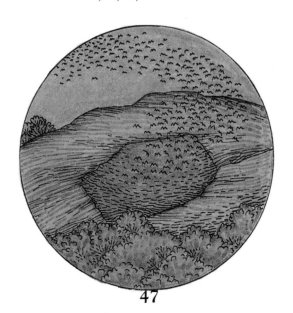

The dark area on this map shows where caves can be found around the world. X indicates where to find a cave like the one in this book.

Find Out More

To find out more, write to or call:
The National Speleological Society
Cave Avenue
Huntsville, AL 35810
205-852-1300
Ask for the **Junior Speleological Society.**

For a list of U.S. caves that are open to the public:
Gurnee Guide to American Caves, NCA Books,
Route 9, Box 106, McMinnville, TN 37110

Further Reading

Look for the following in a library or bookstore:

Golden Guides, Golden Press, New York, NY

Golden Field Guides, Golden Press, New York, NY

The Audubon Society Beginner Guides, Random House, New York, NY

The Audubon Society Field Guides, Alfred A. Knopf, New York, NY

The Peterson Field Guides, Houghton Mifflin Co., Boston, MA

Reader's Digest North American Wildlife, Reader's Digest, Pleasantville, NY